Information Systems Engineering Library

# Reverse Engineering – An Overview

Richard West

CCTA

November 1993

LONDON: HMSO

© Crown Copyright 1993

Applications for reproduction should be made to HMSO

First published 1993

ISBN 0 11 330602 4

For further information regarding this publication and other CCTA products please contact:

CCTA Library
Riverwalk House
157-161 Millbank
London SW1P 4RT

071-217-3331

# Contents

**Foreword**

**Acknowledgements**

**1  Introduction**                                                             1

    1.1    Background

    1.2    Purpose

    1.3    Who should read this volume

    1.4    Assumed knowledge

    1.5    Structure of this volume

**2  What is reverse engineering ?**                                            5

    2.1    Definition

    2.2    Scope of reverse engineering

    2.3    Use of reverse engineering

    2.4    Tools to assist reverse engineering

    2.5    The processes of reverse engineering

    2.6    Associated information

**3  Relationship of reverse engineering with other techniques**                11

    3.1    Software engineering

    3.2    Forward engineering

    3.3    Software maintenance

    3.4    Code analysis

    3.5    Redocumentation

    3.6    Design recovery

    3.7    Restructure

    3.8    Reuse

    3.9    Re-engineering

| | | | |
|---|---|---|---|
| **4** | **Benefits of reverse engineering** | | **23** |
| | 4.1 | Investment in current systems | |
| | 4.2 | Time savings | |
| | 4.3 | Flexibility to change the functionality | |
| | 4.4 | Standards/technical | |
| | 4.5 | Defensive maintenance and minimum fixes | |
| | 4.6 | Quality/audit | |
| | 4.7 | Risk assessment | |
| **5** | **Limitations of reverse engineering** | | **33** |
| | 5.1 | Conceptual limitations | |
| | 5.2 | Practical limitations | |
| | 5.3 | Management issues | |
| **6** | **Producing a business case for reverse engineering** | | **37** |
| | 6.1 | Providing a short-term solution | |
| | 6.2 | Providing a long-term solution | |
| | 6.3 | General issues | |
| | 6.4 | Risk reduction | |
| **7** | **Recommendations for reverse engineering projects** | | **59** |
| | 7.1 | Planning | |
| | 7.2 | Change control | |
| | 7.3 | Metrics and measurement | |
| | 7.4 | Testing | |
| | 7.5 | Pilot projects | |

| | | | |
|---|---|---|---|
| **8** | **Future trends in reverse engineering** | | **61** |
| | 8.1 | Technical developments | |
| | 8.2 | Management issues | |
| **Annexes** | | | **67** |
| | A | Examples of reverse engineering projects | |
| | B | Examples of cost improvement through reverse engineering | |
| | C | Reasons for reverse engineering | |
| | D | Reverse engineering methods, tool types and tool selection | |
| | E | Contracting out of reverse engineering services | |
| **Bibliography** | | | **87** |
| **Glossary** | | | **91** |

# Reverse Engineering - An Overview

# Foreword

The **Information Systems Engineering Library** provides guidance on managing and carrying out Information Systems Engineering activities. In the IS life cycle, Information Systems Engineering takes place once the IS strategy has been defined. It is concerned with the development and ongoing improvement of information systems up to the operational stage, when systems become the responsibility of infrastructure management.

The Information Systems Engineering Library builds on guidance in the CCTA IS Guides, particularly set A: *Management and Planning Set* and set B: *Systems Development Set* and complements other CCTA products, in particular the project management method, PRINCE, and the systems analysis and design method, SSADM.

Volumes in the Information Systems Engineering Library are of interest to varying levels of staff from IS directors to IS providers, helping them to improve the quality and productivity of their IS development work. Some volumes in this library should also be of interest to business managers, IS users and those involved in market testing, whose business operations depend on having effective IS support by means of Information Systems Engineering activities.

The Information Systems Engineering Library also complements related CCTA publications, particularly the IT Infrastructure Library for operational issues and the IS Planning Subject Guides for strategic issues.

CCTA welcomes customer views on Information Systems Engineering Library publications. Please send your comments to:

 Customer Services
 Information Systems Engineering Group
 Gildengate House
 Upper Green Lane
 NORWICH
 NR3 1DW

# Acknowledgements

The assistance of Patrick McDonnell under contract to CCTA from the Centre for Software Maintenance Ltd is gratefully acknowledged.

The assistance of Dave O'Neill of the CCTA staff is also gratefully acknowledged.

# 1 Introduction

## 1.1 Background

In recent years reverse engineering has been the subject of many conferences, articles and books. In addition, several suppliers now provide services and tools in this field.

Unfortunately the term 'reverse engineering' has been used to cover a wide range of activities, with consequent confusion and misunderstanding. This confusion has sometimes led to unrealistic expectations of reverse engineering, or its rejection in circumstances where it could have been of value. Therefore, there is a need to clarify what reverse engineering is, and to provide guidance, based on up-to-date knowledge and experience, of how reverse engineering techniques can be best applied to existing software.

Reverse engineering is defined as:

> "*the process of analysing a (software) subject to identify the system's components and their inter-relationships, and to create representations of the system in another form or at higher levels of abstraction*". (Chikofsky and Cross)

Essential factors in this definition are:

- identification of the system's software components and their internal functions

- identification of the inter-relationships between the individual identified software components and also to the operating environment.

These give insight into the design and structure of the application software. This is discussed further in Chapter 2.

The creation of representations of the system in another form, or at a higher level of abstraction, recognises that modern techniques of software engineering can yield benefits in efficiency and maintainability. Systems implemented using old languages and undisciplined methods can be investigated using reverse engineering. Often, when the detail of the existing system is understood, a more efficient and maintainable structure can be designed, using modern languages and tools. This is the subject of Chapter 3.

Properly used in appropriate circumstances, reverse engineering leads to improvements in the quality of application software supporting an organisation's business needs, through:

- improved reliability and quality of software maintenance

- increased productivity of the maintenance staff.

## 1.2 Purpose

The purpose of this volume is to provide information on the definition, scope and purpose of reverse engineering. The volume also gives practical advice and guidance which can be used to identify where reverse engineering is likely to provide a successful and cost-effective approach to the maintenance of application software systems.

Terms used in association with reverse engineering are explained, and a framework is given to allow readers to understand the relationship between reverse engineering and associated activities.

Although many of the techniques described in this volume are applicable to real-time applications, where timing and interrupts are critical, such systems are not covered explicitly in this volume.

Chapter 1
Introduction

| | | |
|---|---|---|
| 1.3 | **Who should read this volume** | This volume is intended for IS managers and customers of IS services, who wish to identify how and when it would be beneficial to apply reverse engineering to existing software systems. This volume is also likely to be of interest to IS service providers who may need or want to apply reverse engineering techniques. |
| 1.4 | **Assumed knowledge** | No prior knowledge of reverse engineering is assumed. However, a general understanding of the context and scope of software maintenance and project management would be useful. Information on these topics may be found respectively in the ISE Library volume: *Management of Software Maintenance* and the Programme and Project Management Library volume: *PRINCE – A Management Outline*. Further detail of many aspects of these topics may be found in the publications listed in the Bibliography. |
| 1.5 | **Structure of this volume** | The definition of reverse engineering, its scope, nature and purpose are discussed in Chapter 2. |

A range of techniques and activities associated with software engineering and their relationship to reverse engineering, are covered in Chapter 3.

The potential benefits of reverse engineering are covered in Chapter 4.

The limitations in the applicability of reverse engineering are dealt with in Chapter 5.

The factors to be considered when evaluating the case for applying reverse engineering to a given application software system are set out in Chapter 6.

Recommendations for possible reverse engineering projects are given in Chapter 7.

Some likely future trends are discussed in Chapter 8.

Further detailed information is given in the Annexes, covering:

- examples of typical reverse engineering projects
- examples of cost improvements following reverse engineering
- a reverse engineering checklist
- reverse engineering tool types and tool selection
- contracting out of services and reverse engineering.

There is also a Bibliography and a Glossary of the terms used in this volume.

# 2  What is reverse engineering?

## 2.1 Definition

Reverse engineering covers the examination and understanding of existing systems, such as business processes, a software system or a hardware system. The results of this exercise are recorded in a usable or reusable form. Reverse engineering is not a process of change or replication.

The IEEE Standard Glossary of Software Engineering Terminology (Std 601.12-1990) does not contain a definition of 'reverse engineering'. A widely quoted definition, which has gained general acceptance, is that of Chikofsky and Cross (1990). This definition is used in the present volume:

> *"Reverse engineering is the process of analysing a (software) subject to identify the system's components and their inter-relationships, and to create representations of the system in another form or at higher levels of abstraction."*

## 2.2 Scope of reverse engineering

Reverse engineering techniques may be applied widely in the examination of business processes outside the computer, as well as to computer systems that support the business functions. However, this volume only covers the reverse engineering of the application software used to support business systems.

Because the requirements of a business change, the application systems supporting that business must also change. To apply the changes correctly and efficiently, the business processes supported must be understood. Also the sequence of operation of application modules, and their precise function, must be determined. These factors govern the scope for change.

However, past changes to application software may not be well understood. Proper change management procedures may not have been applied. The revised system may give the desired end result; but the way in which changes were carried out may have been undisciplined and the changes poorly documented, if at all. Also the people who originally required the change may, or may not, be available for consultation. These restrictions limit the scope for further change.

**2.3    Use of reverse engineering**

Reverse engineering is a technique which can be used to counteract these effects. It helps software engineers to:

- understand the overall structure and design of an application system

- evaluate the impact of future changes

- identify where files are accessed

- identify the structure and content of the data files

- identify the use of common modules, for example date routines

- examine the detailed code structure

- analyse the quality of the code

- examine the operational details, for example program sequence, parameter input and backups

- document and display the information in a way that is easy to use and manipulate.

## 2.4 Tools to assist reverse engineering

There are many different tools and techniques to assist in obtaining information by reverse engineering. The tools and techniques chosen for a specific task will depend on a number of factors such as:

- the desired level at which information is to be extracted from the existing application software system

- how the information is to be stored (on paper or in a machine-readable form)

- what form the information is to take (design, documentation or code)

- how the information is to be used.

In all cases there are four steps in the reverse engineering of an application software system:

1. Collecting together all the relevant inputs for the reverse engineering task. These are: business requirements, computer system designs, file layouts, code, output reports and Job Control Language (JCL) statements

2. Identifying all the independent objects described in the input

3. Identifying and recording the relationships between the independent objects, together with the conditions which affect the way they are processed

4. Produce information on the current system based on the outputs at 3 above.

If the software subject being reverse engineered is substantial, the use of a repository of documentation and other evidence of the system's behaviour, will be essential. In further development, CASE tools may offer benefit. The precise nature of the outputs from step 4 will depend on the ultimate goals of the reverse engineering exercise. Chapter 6 covers these issues in more detail.

## 2.5 The processes of reverse engineering

Figure 1 is a schematic illustration of the processes involved in the reverse engineering of an existing application software system. The figure shows the information that can be produced, and where that information is used to support other activities such as software maintenance.

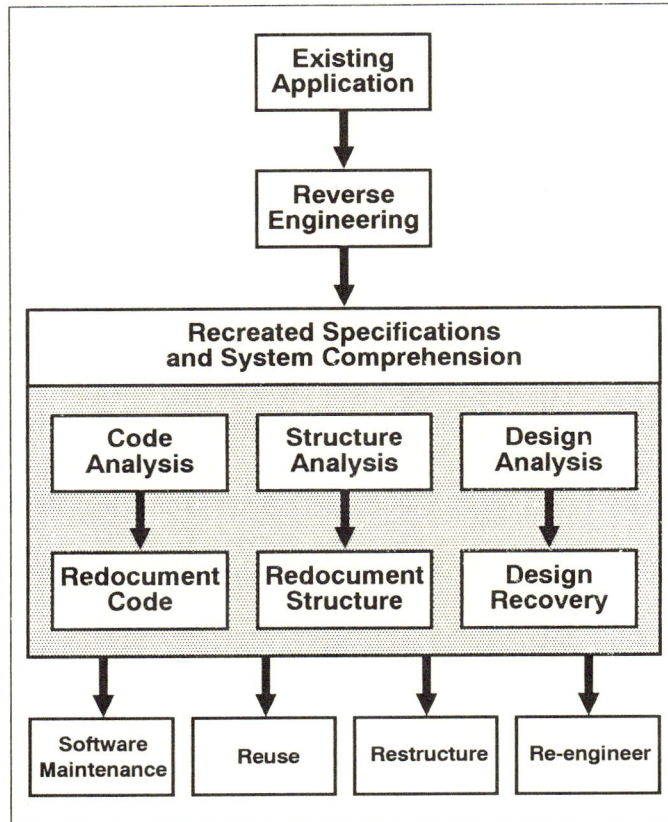

*Figure 1: Reverse engineering schematic*

Reverse engineering can only provide information which is as good as the application system being analysed. It will map out the 'how' but not the 'why' of the IS support to the business process. Thus, if the code does not reflect the correct business process, there is no way that reverse engineering alone can identify or correct that fault.

## 2.6 Associated information

For further details on matters that relate to reverse engineering please refer to:

- the Appraisal and Evaluation Library volumes: *CASE Tools* and *IT Infrastructure Support Tools*

- the Information Systems Engineering Library volumes: *CASE and the Issues for IS Management*, *Improving the Maintainability of Software* and *Management of Software Maintenance*

- the IT Infrastructure Library (ITIL) volumes: *Capacity Management*, *Change Management*, *Configuration Management*, *Help Desk*, *Problem Management*, *Software Control and Distribution*, *Software Lifecycle Support* and *Testing an IT Service for Operational Use*.

…

# 3 Relationship of reverse engineering with other techniques

When reverse engineering tools and methods are used to analyse program code and systems, other techniques may subsequently be employed to provide the total desired solution. The term reverse engineering is sometimes (mistakenly) used to describe the combination of reverse engineering and other techniques, or even to refer to those other techniques. There is a need to identify associated techniques separately and explain their relationship to reverse engineering. This is the subject of sections 3.1 to 3.9.

## 3.1 Software engineering

Software engineering is a generic term covering the development, operation and maintenance of software systems. Reverse engineering and the other techniques discussed in sections 3.2 to 3.9 are all part of software engineering. Figure 2 (overleaf) shows the schematic relationship between the activities.

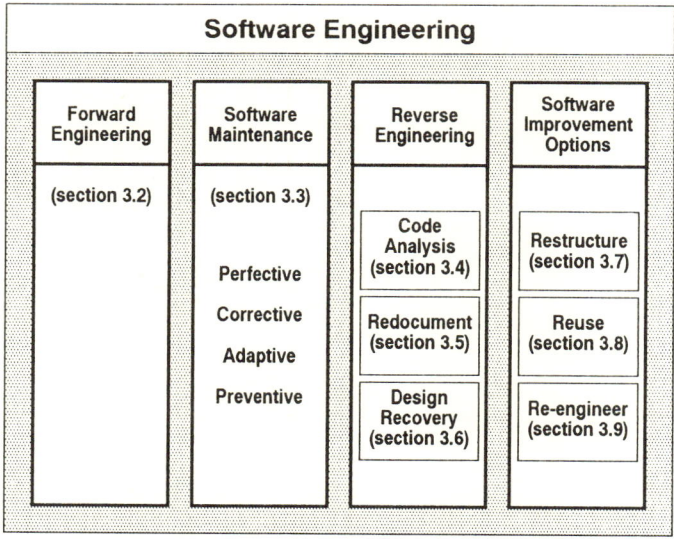

*Figure 2: Relationship of software engineering and other techniques*

**3.2  Forward engineering**

The term forward engineering is used to describe the process of building a new application system from the initial definition of business requirements, through to implementation and live running. Forward engineering is frequently referred to as software development.

Reverse engineering may be involved if the new application system is to be based in part, or wholly, on an existing application system. The details of the new system, for the purpose of accuracy, need to be taken from the operational code of the existing system.

# Chapter 3
Relationship of reverse engineering with other techniques

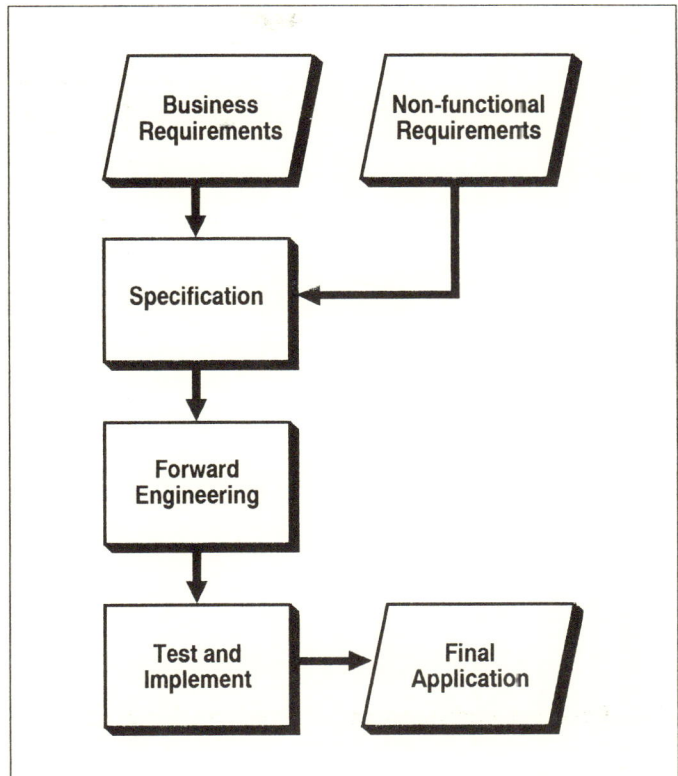

*Figure 3: Forward engineering*

Figure 3 illustrates the typical processes associated with forward engineering. Reverse engineering can contribute to both sets of requirements.

An understanding of the business processes in the existing application systems is necessary to prepare a full specification of the new business requirements. The use of reverse engineering aids this understanding. Analysis of the operational environment and interfaces with other application systems may be a critical task in defining non-functional requirements that are not part of the business process, but are needed to support the business process.

## 3.3 Software maintenance

The IEEE Standard Glossary defines software maintenance as:

> *"The process of modifying a software system or component after delivery to correct faults, improve performance or other attributes, or adapt to a changed environment."* (P1219)

Software maintenance can be classified into four types:

- *perfective*: any modification or enhancement of the existing functionality or performance of application software

- *corrective*: the correction of processing, performance or implementation problems in application software

- *adaptive*: the changes made to application software to adapt it for a change of the supporting environment, network or hardware platform

- *preventive*: action taken to make subsequent maintenance of application software more efficient and reliable.

To carry out any of these types of software maintenance there is a need to understand the detailed aspects of the application system prior to making changes to the software. Sometimes this understanding can be obtained directly from the code or documentation. If not, then the detailed information from reverse engineering may be needed to understand the complexities of the application system and then to specify the changes.

For more information on software maintenance see the Information Systems Engineering Library volume: *Management of Software Maintenance*.

Figure 4 gives an outline illustration of the process of using reverse engineering in both change specification and software maintenance (changing the code).

# Chapter 3
## Relationship of reverse engineering with other techniques

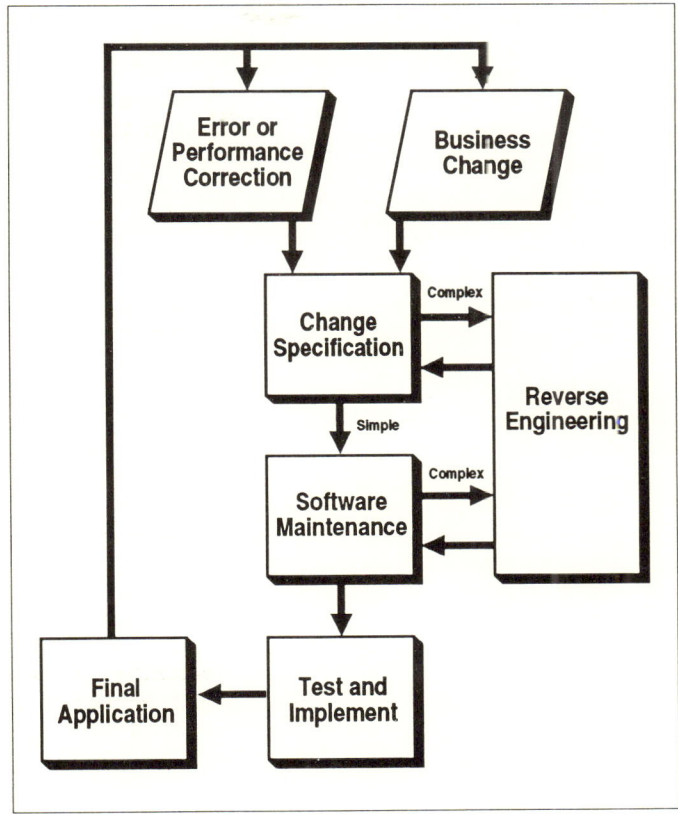

*Figure 4: Software maintenance*

**3.4 Code analysis**

Code analysis is part of, and provided by, reverse engineering (see Figures 1 and 2). A number of tools are available as code analysers only, which produce information on:

- code quality
- impact of a proposed change
- code anomalies.

15

These tools are not as comprehensive as reverse engineering tools, and do not provide redocumentation.

See Annex D for further detail.

## 3.5 Redocumentation

This is the term given to the process of recreating the documentation from examination of the code of the application and other input sources. It can be used to recreate the user, system, design and/or operations documentation. However, there is no standard definition of the types or detail of documentation required and consequently this action may be very varied from installation to installation.

Unless the task is small or very specific, reverse engineering tools and techniques are required. Some tools automatically produce documentation from the code, or use the information extracted to populate a repository from which a CASE tool can produce the required documentation.

The task of redocumentation alone does not normally include the documentation of intended changes or modifications to the application software. If redocumentation is to be carried out together with a project that does lead to change, such as re-engineering, then it is usually considered to be a part of that project.

Figure 1 illustrates redocumentation as one of the products of reverse engineering.

## 3.6 Design recovery

When the design details of an existing application system have been lost or are out of date, it may be quicker, less costly and more accurate to reverse engineer the code back to the design than to work from the original documentation. Design recovery is part of reverse engineering (see Figure 1) and is the task of recreating the overall system design and documentation. Design recovery is a level above documentation recovery for code and files.

# Chapter 3
## Relationship of reverse engineering with other techniques

As stated in Chapter 2, design recovery by means of reverse engineering will only produce the 'how' not the 'why' of the existing design. However, an experienced software engineer who has an understanding of the business process may be able to deduce the 'why'. In this case, the deductions should always be agreed with the business managers or, if available, the original developers.

**3.7  Restructure**

This term is used to describe the process of making the structure of the application software less complicated, more robust, more logical and easier to follow. It is usually carried out at the program or module level. Restructuring does not itself involve functional change, although in practice, the need for it is often a result of functional change.

This activity is basically a technical process which can be carried out by the IS provider. However, when the exercise includes the removal of redundant code, the business managers of the application system must be consulted to ensure that only code which is truly redundant is removed.

The principal objective of restructuring an application is to make it more maintainable; for example, by creating a more logical flow, or bringing the code into line with modern forward engineering methods such as SDM. Reverse engineering techniques and code analysers would be used to identify where restructuring would be of most benefit.

Figure 5 illustrates restructuring, based on the use of knowledge obtained by reverse engineering, to make it possible to accommodate new requirements into an existing application.

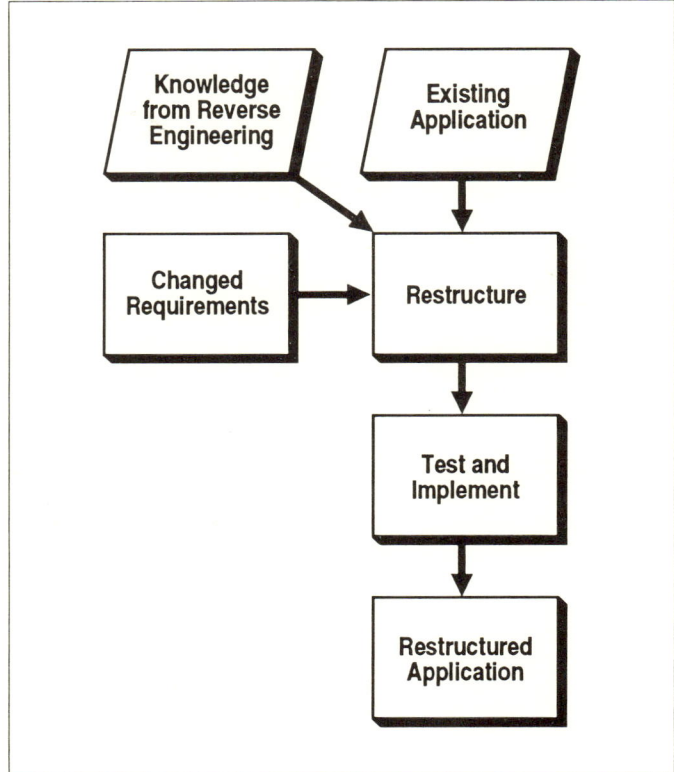

*Figure 5: Restructure*

Restructuring can be used as a tool in preventive software maintenance.

# Chapter 3
## Relationship of reverse engineering with other techniques

### 3.8 Reuse

The term 'reuse' describes taking a discrete part of the existing application system forward to a new application system, or the incorporation of it in another application system. Typical examples of this are the reuse of the code of a lengthy calculation or piece of logic whose function is required in new application software without significant change. In such cases, the existing code can be reused.

Figure 6 outlines the steps for reuse.

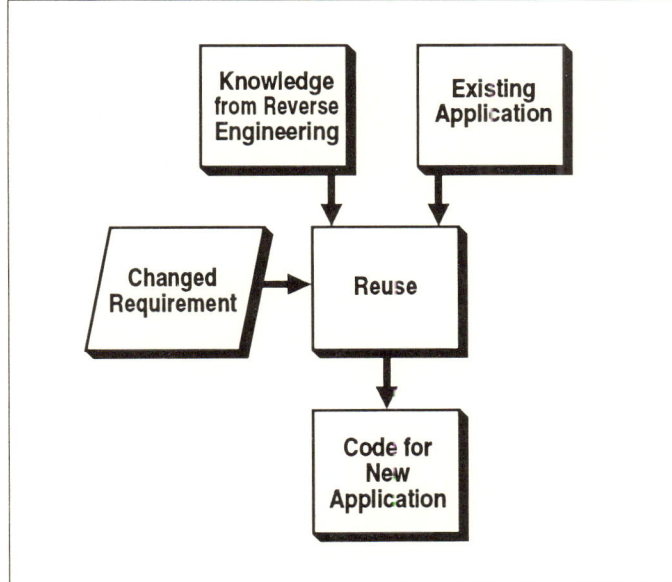

*Figure 6: Reuse*

Although the code is being reused, some changes may still be required; for example to generalise the code or to align it with the latest naming standards.

Reusing existing code has clear productivity benefits. However, where code is used by two or more application systems, there is an obvious need for strict control and documentation. Without control, maintenance problems can arise when the requirements of one application system change while those of the others remain stable.

Reverse engineering is used to assess the feasibility of reuse, and to check the part to be reused for adherence to standards.

## 3.9 Re-engineering

Major changes to business processes often build on or incorporate a significant number of existing procedures. In these cases benefits in time and cost can be gained by re-engineering the existing application code. This re-engineering involves examining the existing application system, by reverse engineering and upgrading with additional or changed requirements. This whole process will produce a new application software implementation economically, to meet both the new and unchanged business requirements.

In some cases the existing application software system may be unreliable, fragile and needing significant support, even though few user changes are requested. It may then be advantageous to reverse engineer the application system and, from the information gained, to re-engineer a more reliable and robust implementation, even though the overall functionality remains unchanged.

Figure 7 illustrates re-engineering.

# Chapter 3
## Relationship of reverse engineering with other techniques

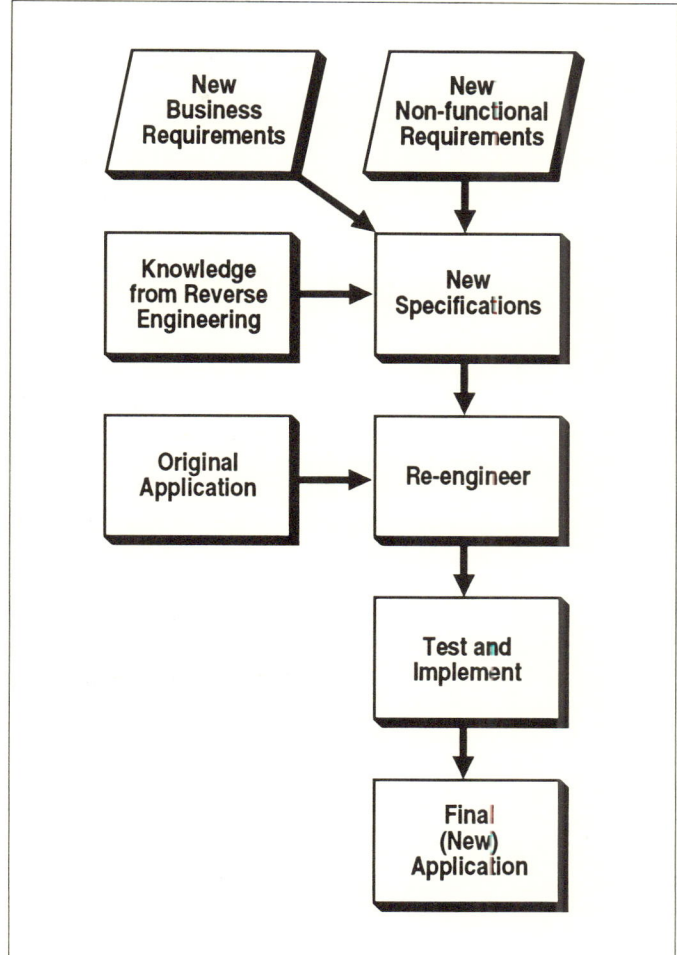

*Figure 7: Re-engineering*

Re-engineering can take place at several levels of abstraction, for example module re-engineering, data re-engineering, design re-engineering, business process re-engineering.

In some cases the term 're-engineering' is used to define the scope of a project which includes both reverse engineering and re-engineering. Re-engineering techniques can be used within perfective, adaptive or preventive software maintenance.

Typical examples would include the re-engineering of application software code for major functional change, a change of hardware platform or a change of operating system or database.

As shown in Figure 2, re-engineering is one of three software development options; it may be used in conjunction with restructure and reuse.

# 4 Benefits of reverse engineering

The main benefits of reverse engineering are:

- maximising the return on the existing investment in application software

- improving the productivity of personnel involved in application software maintenance

- reducing the costs of maintaining an organisation's application software portfolio

- reducing risks associated with change to application software by improving understanding and control, thereby making it easier to change the software.

These benefits tend to overlap. This chapter describes areas and activities where reverse engineering can deliver benefits. The benefits are closely tied to the specific historical development and operating conditions of the application software system. Each case must be evaluated separately: the reverse engineering of similar looking systems may not deliver similar benefits.

Only a small number of examples are detailed in this volume. This chapter is, however, intended to illustrate a framework within which the business cases to justify reverse engineering can be built and evaluated.

# Reverse Engineering - An Overview

Reverse engineering is often associated with change. The forces for change affecting software are illustrated in Figure 8.

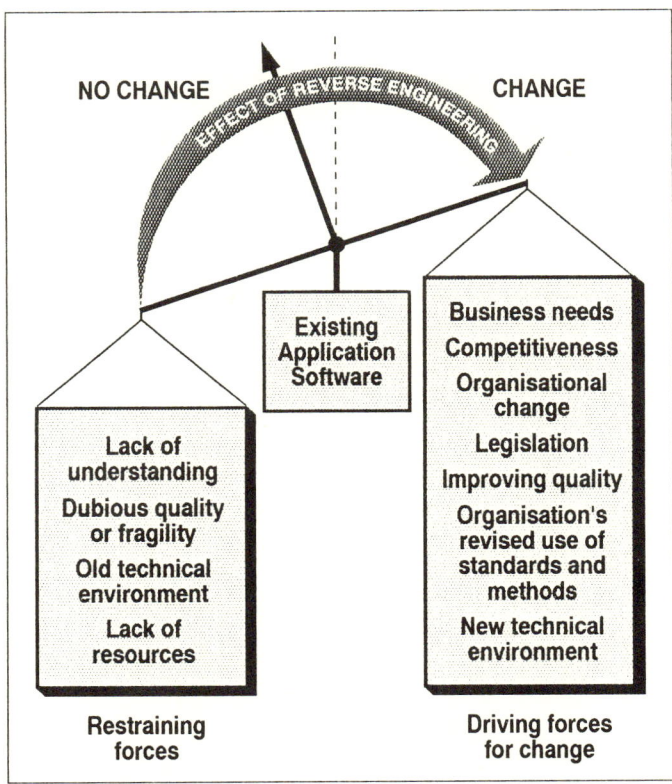

*Figure 8: Forces for change*

### 4.1 Investment in current systems

An application software system in operational use will have required substantial investment, not only in IS development and support, but also in the training of users and in acceptance by customers.

The existing investment in application software includes the cost of specification, development, testing and implementation, maintenance and support. To these must be added the cost of hardware, networks and other items of IT infrastructure.

# Chapter 4
# Benefits of reverse engineering

In most cases such an investment will have produced an application software system which meets most user and business requirements. The use of reverse engineering to extend the life of the system, or to improve the current performance such that it can meet additional or changed demands, can show significant savings over the cost of replacing an old system. There may be greater cost in developing a new system or evaluating and implementing a new package.

If, however, it is necessary to implement a new application software system, the use of reverse engineering approaches can help in the process of changeover.

Most new application software system developments follow on from existing systems, or have to link up with different applications. Although not a part of building the new code, the use of reverse engineering to check existing functionality, or to define interface requirements, may point to opportunities for significant savings in time and cost.

### 4.1.1 Maintenance Costs

The cost of software maintenance during the life-cycle of an application system is very large. A CCTA survey carried out in 1990 at central government sites provided the following information on costs:

| | |
|---|---|
| cost of software development | 35% |
| cost of software maintenance | 40% |
| overheads (including management) | 25% |

The cost of the above software maintenance was distributed as follows:

| | |
|---|---|
| perfective | 58% |
| corrective | 18% |
| adaptive | 17% |
| preventive | 7% |

The use of reverse engineering approaches can enable improvements to be made in the productivity and reliability of both ongoing application software maintenance, and of the re-engineering rather than the redevelopment, of an application.

Reverse engineering can also facilitate the reuse of existing software in development projects, with consequential cost savings.

4.1.2 Productivity

Introducing reverse engineering tools and methods into the labour-intensive activities of application software maintenance and support can show a significant increase in productivity. In addition, it may be possible to provide more flexible support arrangements.

Figure 9 shows the results of a 1991 UK-based survey, by Spikes Cavell, of 365 business (not scientific) mainframe and large mini-computer sites.

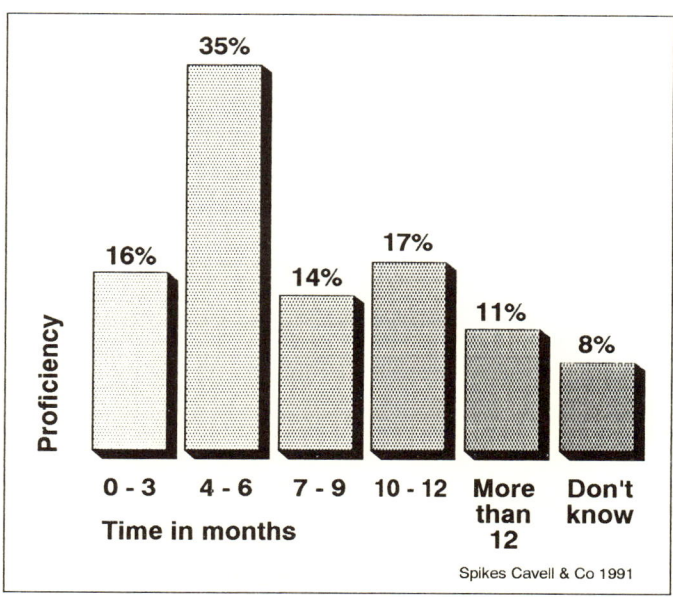

*Figure 9: Average time to proficiency of maintenance staff*

# Chapter 4
## Benefits of reverse engineering

This survey showed that:

- up to 65% of an IT department's staff resources are allocated to the correction of errors, enhancements and migrations

- over 42% of the installations estimated that it took over 6 months before new staff became proficient in software maintenance (11% took over 12 months).

Figure 10 illustrates the breakdown of time spent on software maintenance.

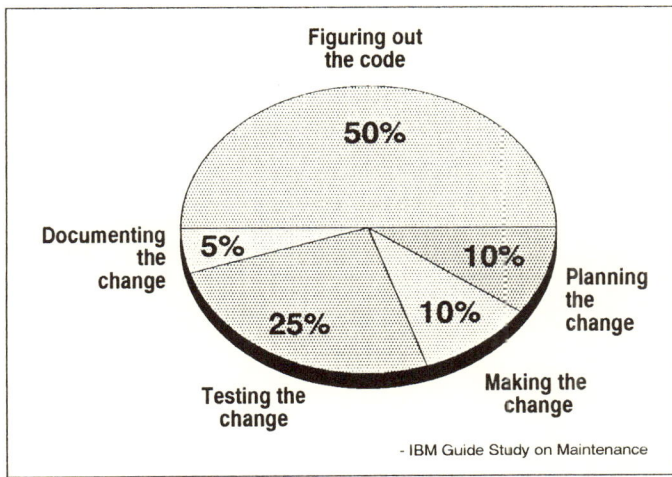

*Figure 10: Breakdown of time spent on maintenance*

Even staff who are proficient in maintaining an application software system can spend up to 50% of their time 'figuring out' the code, particularly if it is unstructured.

Reverse engineering techniques for analysing and manipulating the code can reduce this 50% figure. The improved information, such as documentation and knowledge of the business process gained, will reduce the time required for staff to become proficient in the maintenance of an application software system. Thus, the maintenance team's throughput will be improved.

### 4.1.3 Reliability

Providing a means of analysing the code and checking subsequent work can improve both the reliability of the existing code and that of the amended code. Improved reliability brings down IS provider and customer costs, and results in better services to the IS customer.

### 4.1.4 'Hot spots'

Continuing problem areas in the support and maintenance of particular sections of an application software system are sometimes known as 'hot spots'. These 'hot spots', or areas of unpredictable performance, can add to costs and jeopardise the future of systems. By using reverse engineering techniques, these areas can be more easily identified and analysed, and corrective action can then be recommended.

## 4.2 Time savings

As outlined in Chapter 3, reverse engineering may be the first step in understanding the existing application software system prior to modification, or prior to re-engineering and replacement. In these circumstances the use of reverse engineering may show considerable time savings over the alternative development paths:

- trying to obtain the information without using reverse engineering

- redeveloping afresh

- installing a new package.

Time savings through the use of reverse engineering for system and program comprehension will result from:

- more accurate understanding (particularly if the person analysing the system does not have in-depth knowledge of it)

- reduced user involvement, as there is less need to involve lots of users' time to check detail

# Chapter 4
# Benefits of reverse engineering

- reduced redevelopment timescales - the current functionality is accurately defined

- reduced timescales for package selection and evaluation - the current functionality is accurately defined.

## 4.3 Flexibility to change the functionality

As the business environment and business processes change, supporting systems must mirror them. Reverse engineering can assist in this process of change through:

- accurate analysis of the functionality of the current application

- impact analysis of the effects of change

- analysis of the technical aspects and feasibility of change.

## 4.4 Standards/technical

Systems which have been developed and modified have often been subjected to a variety of standards, methods and procedures or, on the other hand, they may have been developed without any standards at all. For example, the system requirements may not have been analysed using SSADM, or the code may not have been developed in accordance with methods such as SDM. Individuals or teams involved may have used their own ideas of standards, or other local initiatives may have been applied; for example:

- the 'quick fix'

- individual programmers' naming standards

- informal naming standards adopted by work groups.

An essential business system should be developed and maintained in accordance with the organisation's current standards, methods and procedures. Failure to do this may cause significant future problems, particularly as support staff lose familiarity in dealing with systems that do not conform.

Reverse engineering tools and techniques can play a significant part in identifying departures from standards and in checking that the revised standards have been applied consistently and correctly. For example, prior to installing the latest version of a compiler, database or operating system, programs can be examined to identify code which may not be compatible with the new version.

### 4.5 Defensive maintenance and minimum fixes

Some programs suffer from 'defensive maintenance' where programmers – unable fully to understand the existing code – write new, additional code instead of changing the existing code, or even introduce additional files. The net result is to add to the overall size and complexity of a program making it harder to maintain.

Complexity can also be introduced by the 'minimum fix', the requirement that any change should be completed with the minimum amount of code alteration.

Reverse engineering techniques will:

- help interpret complex programs and reduce the risk of programmers introducing additional complexities

- identify complexities introduced in earlier maintenance.

### 4.6 Quality/audit

There are many recommended standardised techniques and methods for producing high quality work, for example SSADM for structured systems analysis and design. Most organisations use one or more of these methods, together with appropriate national or international standards and their own procedures. The methods chosen often depend on the programming language and hardware/software platform used. Standards, methods and procedures ignored or poorly applied could give rise to future problems.

Consideration should be given to using reverse engineering not only to analyse the structure of the code but also to produce statistics which enable the quality of the code to be assessed. Typical statistics include numbers of file and module calls, particular code constructs, for example GOTOs in COBOL and drop-through options.

The quality of software maintenance will be improved if reverse engineering methods and information are available to help with the process.

Although most reverse engineering is applied to older code or heavily modified code, using reverse engineering methods on new code for checking quality and adherence to standards also has benefits. Reverse engineering is not a once and for all activity, but should be a part of steady-state software maintenance to check for adherence to standards.

Annex A gives an example of the use of reverse engineering to check that the quality of code delivered by a third party meets contractual standards.

## 4.7 Risk assessment

The previous sections suggest specific ways in which reverse engineering can help with the process of software and system management. Management includes the continuous assessment and balancing of risk. Reverse engineering can be used to provide information on the basis of which management action can be taken to reduce the risks identified. Typically the actions are as follows:

- identifying potential trouble spots arising from complex or poorly designed systems and poorly written software

- quality checks on new and existing software to identify potential problems

- impact analysis to determine the extent of any changes before they are initiated, to reduce cost and timescale risk

- eliminating or reducing the dependency on staff who have detailed knowledge of the system, but who may not be available when required

- understanding the scope of the business logic in a system to reduce the risk of future misalignment of the system with the business it is supporting

- understanding the technical details, such as the extent to which reusable code modules for common processing are used, to reduce the risks of accidental future changes.

# 5 Limitations of reverse engineering

The technology of reverse engineering has, in some cases, been oversold with the consequent possibility of projects failing and causing risk to the organisation.

Sometimes this has led to a rejection or mistrust of reverse engineering.

There are conceptual and practical limitations to reverse engineering together with management control issues. These are covered in sections 5.1 to 5.3.

## 5.1 Conceptual limitations

Reverse engineering can only provide information about how an application system works. It cannot produce the information about why it works as it does. This is a significant limitation for four reasons:

- the 'why' information is needed in order to make informed changes to a system

- capturing the 'why' information is an expensive knowledge-based task, which should be undertaken in conjunction with reverse engineering

- reverse engineering will represent the code constructs, but cannot distinguish whether the sequences (A before B before C) meet a precise requirement, or happened to be an arbitrary choice by the analyst or programmer which can now be rearranged

- reverse engineering cannot recover the reasoning behind any need for variable parameters.

However, in conjunction with the reverse engineering outputs, an experienced software engineer can in many cases make accurate assumptions on the 'why'. Confirmation by the business managers or, if available, the original users or developers, is then a relatively simple and short task.

## 5.2 Practical limitations

Before proceeding with reverse engineering, detailed checks must be made to ensure that:

- exploitation is made of any appropriate reverse engineering features that are already embedded in other software tools, such as library handling routines and compilers. Checks should be made on the prerequisites for using the tool and on the minimum software environment needed to support its use

- the reverse engineering software tools being considered will run on the precise target configuration, and will produce all the analyses required. Consultation with other users of the tools should help

- significant benefit can be obtained by using reverse engineering

- output can be stored and accessed in a way that is entirely practical. For example, a check should be made that the output does not need special conversion from tape to 3.5" disk before it can be used on a PC. Such a necessity would severely limit progress on tasks which require multiple analysis. Again, knowledge of other users' experiences can be invaluable

- the use of the tool does not require special embedded code in the applications being analysed.

Also, before proceeding with a reverse engineering project, always consider a pilot evaluation. Acceptance of associated tool(s) will depend on a successful outcome (see Chapter 7).

# Chapter 5
## Limitations of reverse engineering

For further detailed advice related to selecting CASE tools to support reverse engineering, please refer to:

- the Appraisal and Evaluation Library volumes: *CASE Tools* and *IT Infrastructure Support Tools*

- the Information Systems Engineering Library volume: *CASE and the Issues for IS Management*.

### 5.3 Management issues

The issues which may need addressing to maximise the benefits of reverse engineering include the following:

- the need to apply reverse engineering may indicate that, at some point in the past, the management of the IT function has been lax. In a worst case scenario, it may be that there have been ineffective procedures covering change control, and the whole system infrastructure (such as system libraries and versions of the code used) could be suspect. This possibility should be checked and if the problem is verified, an approach should be agreed to minimise any immediate implications and to correct the situation. The references at the end of Chapter 2 may be helpful

- although it goes a long way to identifying what is being used and where, reverse engineering cannot solve problems of discipline and procedure

- unless the issues identified from the reverse engineering exercise are assessed and, if necessary, positive managerial steps are taken to avoid recurrence, there will be little if any long-term benefit

- use of reverse engineering can raise many technical issues. Unless managers of projects exploiting reverse engineering clearly define the objectives and goals, and monitor progress towards their achievement, there is a risk that these objectives will not be met

- operating systems, databases, report generators and application package code may be protected by legal agreement. The use of reverse engineering tools for interrogation of these items of software may violate these agreements (see section 6.3.5)

- software (as above) may be licensed to specific machines on specific sites. Even if technically feasible, reverse engineering must not be used to assist in the transfer of software to other machines or sites without any prior consent that may be required

- the proposed project must abide by the Copyright (Computer Programs) Regulations 1992 (see section 6.3.5)

- a successful reverse engineering project may only confirm, in a technical sense, that which was perceived from a managerial or a maintainer's view – what was thought to be rubbish is proven to be rubbish!

For further detailed advice on management issues related to CASE tools which are used for reverse engineering, please refer to the Information Systems Engineering Library volume: *CASE and the Issues for IS Management*.

# 6 Producing a business case for reverse engineering

The previous chapters have identified the opportunities and the misunderstandings that are associated with reverse engineering. If, after clarifying these it seems that reverse engineering may provide a means of achieving the desired goal, then a more detailed evaluation must be completed.

This chapter outlines the topics, and lists both the technical and non-technical issues to be considered when producing a business case for using reverse engineering on a project. The solutions to problems that reverse engineering may provide are both short term and long term; both types of solution have potential costs and savings. There are also issues that affect both types of solution. This chapter addresses the issues in the following order:

- short-term solutions (section 6.1)
- long-term solutions (section 6.2)
- general issues (section 6.3).

The chapter concludes with a section on risk reduction.

## 6.1 Providing a short-term solution

For the purposes of this volume, a short-term solution is the use of reverse engineering to address an immediate problem, such as corrective, perfective or preventive maintenance, or its use in gaining information as input to the analysis phase of system redevelopment. Using reverse engineering for short-term solutions does not, in general, lengthen the life of an application system, enable the system to be transferred to different hardware, or have wider implications elsewhere in the organisation.

The decision-making process, when seeking justification for obtaining a reverse engineering capability, must be rigorous. The capability may be provided by software, tools, documentation, procedures and training; all of these cost. The factors triggering the decision are usually internal to the support teams. The justification can be in terms of time saving (and, therefore, cost saving), flexibility in the application of future changes to the application software system and quality of work. For all application software systems to which the reverse engineering capability is to be applied, these advantages must be balanced against the cost of acquiring that capability, and using it.

The justification for using reverse engineering in analysis for the development of a new application will be based on factors such as accuracy, time saved and comprehension. These are to be set against the costs of alternative methods, such as manual code analysis, user interviews and relying on existing documentation.

For short-term solutions, the costs and savings to be considered in the decision process are:

*Costs*

- the cost of identifying the requirement for a reverse engineering capability, and the cost of evaluation of an offering prior to purchase

- the cost of purchase/rental and any on-going maintenance charges for software tools, procedures and/or documentation which comprise the reverse engineering set

- any additional hardware costs, for example high resolution screens

- the cost of training, including lost staff resource while the staff are trained.

*Savings*
- a reduction in time spent analysing the code and the system, including the time of users who have to provide information about the existing system

- a reduction in the time spent making changes

- a reduction in the time spent in testing. This can be significant, and is likely to involve both the existing and upgraded systems

- code which requires less support and is easier to change in future, which saves:

    - resources required to investigate and correct errors

    - call out payments if appropriate

    - user time in error identification and correction

    - consequential loss of service to the clients

    - operational and machine time resources to recover/rerun

- intangible savings such as fewer queries from users, and greater use of the system facilities by the users when they come to trust it.

Any staff resources released as a result of reverse engineering are usually highly skilled in technical and business process areas.

## 6.2 Providing a long-term solution

Using reverse engineering to achieve a long-term solution, such as transferring the application system to run on the latest hardware, or re-engineering to align with the current (and future) business processes, can show significant savings of cost and time when compared with alternatives, such as redevelopment.

Typically, the systems an organisation inherits will have been developed and maintained over many years, by a large number of people, using a variety of standards, methods, techniques and styles. This means that maintenance staff frequently have to work on application software systems which differ from one another considerably. Not only is it likely that there will be differences in the development approach used for those systems; the systems may also contain enhancements written to different standards from those used in the original development.

Reverse engineering, leading to re-engineering, offers a practical and cost-effective approach to bringing all these software systems into the same analysis, design and implementation environment as is currently used for new development. This would have the benefit of reducing training for maintenance staff, because they would not need to be familiar with the previous discarded approaches. It would also lead to increased productivity, and contribute to the long term improvement in quality of software maintenance, and to the overall quality of the software itself.

The wider and long-term issues have to be considered, including the need to check that:

- the quality of the application system, and its flexibility, will match the future business process requirements

- the business benefit provided outweighs the cost of its provision

# Chapter 6
## Producing a business case for reverse engineering

- the demands placed on the technical and operational environments to support the application software system can be accommodated. These will largely be based on the concepts and design criteria, dating from several years earlier, of the system to be reverse engineered.

The checks to be made will include those discussed in the following sub-sections.

### 6.2.1 Business process

Any reverse engineering activity should be used to improve the way in which the application systems support the overall business process during their likely lifespan. Particular issues to be considered are:

- are the market/service conditions such that there is likely to be a continuing need for application software systems like those being considered?

- is the lifespan of the particular business process likely to justify the cost and effort?

- how volatile will the business be, and are the current application software systems of sufficient quality and flexibility to cope?

- will the application software systems match the hierarchical and geographical organisation of the business?

- are there other IT related business issues to be considered, for example, adding EDI and imaging facilities?

### 6.2.2 Hardware and networks

The expected life of the current supporting hardware and networks (or of probable future replacement) should not be less than the anticipated life of the reverse engineered application software systems.

### 6.2.3 Operating software

Unless the technical environment is supported into the future, reverse engineering and using the information obtained to rebuild (re-engineer) the current application software, for the current technical environment, will be of no long-term use.

Consideration must be given to:

- operating systems
- compilers and languages
- databases
- transaction processing
- report writers
- software interfaces
- network software
- operational schedulers.

Also, will the system require additional translation programs (emulators) to enable it to run on the new hardware?

### 6.2.4 Application software standards

The advantages of reverse engineering and re-engineering may be lost if the application software that has been re-engineered does not adhere to the organisation's current use of standards, methods and procedures. Also, it must be sufficiently flexible to adhere to anticipated future standards.

Standards may cover items such as:

- program structure
- naming conventions
- use of common modules
- use of CASE tools/repositories

- file usage and conventions

- database usage and structure

- system and user security

- version control

- the organisation's style guides, covering, for example, screen look and feel issues, report layouts

- documentation structure and presentation.

6.2.5 Third party application software

As a general rule, a third party application software system that is sold in the market as a package is sold under an agreement which prohibits reverse engineering. In these cases, permission will be required before reverse engineering can be applied.

An organisation which has commissioned bespoke application software code developed by a third party, usually has ownership rights which allow reverse engineering without further agreement. However, some systems have been developed as 'hybrids'. For example, the main application software code may be bespoke but report writing and enquiries may be dealt with by package code.

In all cases the ownership and rights to the software should be checked as an initial step.

Other checks to be made are similar to those for the hardware. Questions to ask are:

- will the software continue to be supported?

- will the software run on the target hardware?

- if the software is run on larger or more powerful machines, will additional licence fees be payable?

Reverse Engineering - An Overview

- are the interface requirements the same?
- is the software part of the organisation's strategic software portfolio?
- will the contractual terms change?

6.2.6 Tools and repositories

As indicated in section 6.2.4, the advantages of reverse engineering may be lost if the software does not adhere to the organisation's current standards.

The organisation may wish to exploit software tools:

- CASE tools
- repositories
- testing tools
- debugging aids
- documentation tools.

Organisational standards may govern the selection and use of such tools.

Checks should be made that the use of reverse engineering on the application software in question is compatible with the use of these software tools, and the standards associated with them.

Further information and advice on this area may be found in the Appraisal and Evaluation Library volumes: *CASE Tools* and *IT Infrastructure Support Tools,* and from the Information Systems Engineering Library volume: *CASE and the Issues for IS Management.*

6.2.7 Alternative solutions

The business case for using reverse engineering approaches prior to re-engineering must consider any alternative solutions. These alternative non-reverse engineering solutions are likely to include those discussed below.

# Chapter 6
# Producing a business case for reverse engineering

*Do nothing*  In the context of this volume, 'do nothing' means do not reverse or re-engineer, redevelop or purchase a new package. The questions to answer are:

- is the software 'mission critical' or can the service demands be reduced to eliminate the need for change?

- can the schedules be revised, for example, to daytime running to enable more support to be given from experienced staff?

Overall, on the basis of the 80/20 rule, are there any other limited actions which can deliver maximum benefit?

*Redevelop*  Total redevelopment from scratch of entirely new application software, without reference to the existing application system, can be attractive, particularly to some of the staff involved. However, if reverse engineering is not used to extract relevant information from the existing application system new risks can be introduced, such as the following:

- 'wish lists' may be generated by business managers, and technical and user staff. Without rigorous control and cost benefit analysis, it will require considerable time and effort to distinguish between the necessary and the 'nice to have' requirements

- the change-over from the old application to an entirely new application system is likely to be more complex

- end user education and training issues may be significant

- the development resources required may be greater or may be harder to estimate

- development timescales may be longer
- higher risks associated with the use of new application software and hardware technology may be brought into play.

*New package*  Many of the management risks associated with redevelopment may apply. There are additional issues to address which may take considerable resource and time:

- package selection
- package modification
- modification of real business needs and practices to the constraints imposed by the 'best fitting' package
- additional staff training and familiarisation for all staff using the package. For example, it may not conform to the usual organisation's look and feel, or other standards
- package support and maintenance
- having to undertake a supplier approval exercise.

Various volumes in the Appraisal and Evaluation Library may be of use in any package selection exercise.

6.2.8 Cost savings and benefits

Any reverse engineering project must, in the final analysis, show benefits outweighing costs. The section will guide the IS manager or 'intelligent customer' through those issues which may influence the cost/benefit equation.

# Chapter 6
## Producing a business case for reverse engineering

The cost benefits analysis of longer term reverse and re-engineering solutions will, by virtue of the fact that the reliability of the code has been improved, include those of reverse engineering as a short-term solution as detailed in section 6.1. Additional costs and benefits to be considered are as described below.

*Costs*

Costs due to:

- hardware requirements to support application software systems. Sharing hardware resources may represent a considerable saving when compared to the costs of an alternative solution. Note that maintenance on older hardware may be expensive

- other software unique to that application software system, and which incurs ongoing support costs, such as:

    - compilers

    - emulators

    - database systems

    - operating system (or different versions)

    - communications software

    - report writers

    - other tools

- enhancing the application software system to meet the organisation's current use of standards, methods and procedures (programming, testing, operational, security and documentation). This is particularly important when obsolete software is involved

47

- retaining particular skills solely for the maintenance and support of the application software system

- the cost to the business of not having a newly developed application system which, through utilising all the latest advances, is able to provide some further business advantage.

*Benefits*

The benefits are:

- improved return on investment

- the maintainability of the application software is improved, thereby reducing costs and timescales for future maintenance, including enhancement

- the end users and clients will receive familiar outputs and the 'learning curve' for transferring to the new system supporting the business application will be eliminated or very greatly reduced

- the 'learning curve' will be lower for both end users and maintainers transferring from other applications due to the use of a common organisational style and underlying standards, methods and procedures

- the costs of changing any links to other systems may be low

- if the system runs on the same machine the additional hardware and operating software costs may be very low

- if there are no significant changes to terminal equipment then the building and cabling costs are likely to be minimal

- if the output format is retained then the stationery and printing costs will be unchanged.

# Chapter 6
## Producing a business case for reverse engineering

## 6.3 General issues

### 6.3.1 Assessing support tools

A key element of the feasibility study is the technical assessment. Applying reverse engineering involves a number of technical stages which comprise different processes from those for previous software maintenance. These processes, and their impact on future maintenance after reverse engineering, should be carefully checked.

Points to be considered are:

- is the reverse engineering a one off activity which will result in re-engineered code that resides in standard libraries, and is incorporated into application software under consistent maintenance and control procedures; or will the reverse engineering techniques have to be used repeatedly prior to every future modification? What is the impact on maintenance productivity and costs?

- is the tool fully interactive or just a code analyser? Interactive tools allow search, navigation and manipulation of the code, whereas analysers present a static view of the existing code. The difference between the types can have a major impact on productivity and timescales

- does the resulting re-engineered application code rely on embedded code from the tool such as a translator, or does it call special modules for execution?

- does the tool have any additional machine resource requirements, for example for memory or dedicated devices?

- will the tool interface with the standard library routines used?

See Annex D, reverse engineering methods, tool types and tool selection.

## 6.3.2 Operational issues

Reverse engineering is often undertaken to extend the life of an application software system by improving the organisation's understanding of the system, and its ability to support that system. However, this will only give limited long-term benefit if the resulting application software system does not fit in with current and planned IT operations. Issues to be considered in deciding whether to use reverse engineering include:

- are there any operational possibilities anticipated that were not originally catered for in the existing application? For example:

    - 'the year 2000 problem', where the original developers had not catered, in the application's code, for the year changing from 1999 to 2000

    - maximum number limits, such as unique reference numbers or file size limits, that may only impinge after many years of successful operation

- is unattended operation required, where application systems run with no central operator intervention whatsoever? Examples of changes required to accommodate this include the replacement of exchangeable tape files with permanently mounted disk-based files, and the printing of reports at remote locations under the control of an operator at these locations

- is the application system migrating, for example, from proprietary to open systems, or is there any significant architectural change?

- have the backup, restart and recovery procedures for the application system been reviewed in line with the latest organisational standards and procedures?

- have the operational Service Level Agreements been agreed?

- what Help Desk training is necessary?

### 6.3.3 Business and user implications

One of the advantages of reverse and re-engineering an application system is a reduction in business managers' and users' involvement. Their time is not wasted on ascertaining the detail that is extracted from the existing application software. Instead their involvement is limited to simply confirming what has been found, and to resolving any outstanding questions, which:

- can shorten the schedules

- avoids 're-inventing the wheel' and creating 'wish lists' which might take the project over its budget.

However, serious consideration should be given to changes in the user requirement, and in the IT facilities available to users, since the original application system was designed. Typical items are:

- the outstanding changes requested by business managers may become technically easier to implement during the re-engineering process, or subsequently

- technical developments made since the existing application software code was developed may potentially be used to supersede parts of the existing application code. For example:

    - the database extract and reporting facilities may now be able to provide many of the original custom-built facilities, and allow an increase in flexibility

    - the application system may have been designed when the user had poor access to terminals, or the use of PCs for local processing was very limited

51

- changes to the organisation's style guides, methods, procedures and use of standards since the existing application software was originally developed. These may require changes in the new system to bring it into line in so far as the business managers, users and software maintainers are concerned, for example:

    - windows and graphical displays not formerly available may make the application system easier to use. This would reduce training costs, increase user efficiency, or add to the value of the system by making possible graphical display and manipulation of information

    - re-engineered application systems can be designed to take advantage of increased user experience, skills or computer literacy, to provide more sophisticated solutions

    - new requirements such as the direct down-loading of files for PCs may replace the original facilities, which may not now be required. For example, data could be made available in a format for directly loading into a spreadsheet, rather than in a printed report. This avoids the need to enter the data by hand into the spreadsheet

    - changes to procedures and methods since the application was originally developed may have a direct or indirect impact. For example, changed organisational procedures, or introducing tools to implement, control and enable automation of future regression testing for all new and re-engineered software, may make future testing during maintenance more efficient for that application.

# Chapter 6
## Producing a business case for reverse engineering

### 6.3.4 Non-technical issues

*End user involvement*

Besides the business managers, the end users also have to be considered, informed and their co-operation and enthusiasm for the reverse and re-engineered systems encouraged. Plans to manage their expectations must be made at an early stage and should be considered as part of the initial evaluation. One way of achieving this, avoiding both the wish list and the 'I would not start from here' syndrome, is to set realistic expectations of the allowable changes. Examples of this are to agree:

- a budget for user sponsored changes

- a minimum threshold for the acceptable cost-to-benefit ratio that any proposed changes have to meet before they are considered

- a timescale limit in which changes have to be agreed

- areas of the system that can or cannot be changed, for example, that database contents cannot be changed, but that output reporting can.

While evaluating suggestions, it should always be borne in mind that end users proposals may be conditioned by their knowledge of the original application system, rather than an understanding of the real business requirements.

Whilst considering the end users, education and training on the new reverse engineered application system should not be overlooked. Since the original implementation, many of the users may have changed and been trained by their colleagues without the advantage of 'official' education. Although the look and feel of the new application system may be identical, a refresher course may be very worthwhile to improve the end user's overall knowledge of the application system and its scope. This training may have a secondary benefit of reducing the number of calls for assistance to management and the Help Desk.

*Personnel/skills*

Despite all the latest tools and techniques, application system development and maintenance is still a labour intensive operation. Although reverse engineering is likely to reduce the time spent on software maintenance, maintenance still requires highly trained staff. Many development staff equate their progress with working on the latest technical innovation.

Alternative ways of measuring technical and career progress have to be devised and communicated to the staff who are due to work on maintaining and re-engineering the older systems. Practical experience suggests that gains of up to 20% in output can be achieved through setting the right targets and motivating staff. Plans must be made and agreed at the earliest stage to involve and motivate the technical staff. Suggestions for this are to:

- emphasise the technical innovation and skill required to be part of a reverse and re-engineering team or in a redevelopment section

- emphasise the management skills and disciplines to be learnt, and required by a project that involves reverse engineering

- set clear and achievable metrics and targets both for short-term and long-term goals, and advertise their attainment

- provide a reward policy, both career and remuneration, and ensure that staff understand it

- establish and encourage formal and informal contact between the reverse and re-engineering team and other development staff

- make commitments to rotate staff between different work areas if they wish, and even more importantly, to make sure that the rotation occurs

# Chapter 6
## Producing a business case for reverse engineering

- make commitments to ensure reverse and re-engineering team staff have the same access to training and education as other staff

- adopt a clear customer/provider approach and establish a good relationship with the customer

- ensure that the cost benefit assessment, prioritisation and change control functions are properly exercised to achieve a realistic work programme.

Unless the personnel issues are addressed, more problems may be created than are solved.

Further information and advice on some aspects of this area may be found in the Information Systems Engineering Library volume: *Management of Software Maintenance*.

*Legal*

The contractual constraints on whether or not an item of application software may be reverse engineered do not necessarily depend only on 'ownership'. The contractual clauses should be reviewed as part of an initial check. The following is a guide to the type of constraints that may apply.

Reverse and re-engineering code may be developed by an organisation that retains the intellectual copyright. If this is done using a standard language such as COBOL and standard manufacturer supplied software utilities, such as database and sort routines, it is unlikely that there will be a problem. However, many of the software utilities supplied by third party specialist developers are often licensed to run on a specific size of machine, on one site and for a particular organisation. These may be system utilities such as backup and restore, other utilities such as report writers, or complete application packages. Changing any of these may contravene the licence and incur increased cost.

Concentrating all the work in one large centre may have many attractions, but not if the licence conditions for a small piece of third party software used in one application suddenly results in a charge for running on a machine configuration fifty times as powerful. Not informing the supplier of the changed platform may be a breach of contract.

Software agreements for products supplied by third parties usually preclude changing the code, particularly of 'package solutions'. Where this is the case, any changes would contravene the agreement and probably any associated maintenance agreements. This does not usually apply to bespoke developed code, but the developer may retain joint intellectual copyright.

If the reverse engineering tool is a long-term investment, then proper procedures to evaluate potential suppliers should be followed. It may be thought that escrow arrangements need to be made, to safeguard the availability of the code of the reverse engineering tool if the supplier ceases trading.

Contractual constraints should be checked if the reverse engineering tool is to be used by staff other than the permanent staff of that installation. For example, the support of the system may be contracted out to third party staff off-site (which may now include cross border use).

Further details of copyright issues may be found in the Copyright (Computer Programs) Regulations 1992 which implement the EC Council Directive on the legal protection of computer programs (91/250/EEC) in the United Kingdom.

When software development and maintenance is contracted out, and when reverse engineering itself is contracted out, further issues arise and need to be addressed.

# Chapter 6
## Producing a business case for reverse engineering

**6.4 Risk reduction**  Many of the risks associated with undertaking a reverse engineering project have already been highlighted. Key areas to be considered are shown in Figure 11 below.

| Situation | Risk |
|---|---|
| Are the users committed to the application system and is it in the mainstream of their activity? | The project may be cancelled part of the way through. |
| Do the strategic business issue still indicate a continuing need for such an application system? | The project may be a technical success but may not fit future business requirements. |
| Are the supporting hardware and networks likely to have a life at least equal to that demanded of the reverse and re-engineered system? | A successful project but application has no hardware on which to run part of the way through its extended life. |
| Has that reverse engineering tool been used in the same environment with the same tools and what were the results? | Knowing exactly why the tool will handle other databases and not the one in use, the project remains unfinished or goes over budget. |
| Will staff still have the skills and motivation to work on this application towards the end of its life? | Expensive maintenance, and few staff are able to support the application software. |

*Figure 11: Key risks*

# Reverse Engineering - An Overview

# 7 Recommendations for reverse engineering projects

The key recommendation is that tasks using reverse engineering approaches are treated as projects, and are set up to reflect the best practices of project management control and reporting, using methods such as PRINCE. In addition, there are further management recommendations on planning the approach and setting the critical success factors. These are as follows.

## 7.1 Planning

Plan to take a lot of small steps rather than one giant leap, preferably changing only one variable at a time. The great value of this is that any problem can be quickly isolated and solved. For example, analysing one file structure will isolate naming inconsistencies, whereas analysing all the files and the logic at the same time will probably end up in confusion.

For further advice on this area, please refer to the Information Systems Guide volume A5: *A Project Manager's Guide*, and the Programme and Project Management Library volume: *PRINCE - A Management Outline*.

## 7.2 Change control

Adopt formal change control procedures. These are critical in any project, and in reverse engineering are vital. Programmers must not be allowed to 'tidy up' or experiment with the code using their own initiative, or if they do so, there must be strict recording and reporting before and after the detailed changes are made.

For further guidance, please refer to the Information Systems Engineering Library volume: *Management of Software Maintenance*, and from the IT Infrastructure Library (ITIL) the volumes *Change Management, Configuration Management, Problem Management* and *Software Control and Distribution*.

| | | |
|---|---|---|
| 7.3 | **Metrics and measurement** | Set realistic metrics covering short time periods (one to three days) within the overall plan. This will identify any project slippage at an early stage and, as a consequence, the longer term schedules are more likely to be achieved. |

For further advice on this area, please refer to the Information Systems Engineering Library volume *Estimating with Mark II Function Point Analysis* and the Programme and Project Management Library volume *PRINCE - A Management Outline*.

| | | |
|---|---|---|
| 7.4 | **Testing** | Testing must be planned and comprehensively evaluated to test all conditions, not just those which have been specified as changes. Regression testing must be planned for. |

For further guidance, please refer to the Information Systems Engineering Library volume *Management of Software Maintenance* and the IT Infrastructure Library volumes *Software Control and Distribution* and *Testing an IT Service for Operational Use*.

| | | |
|---|---|---|
| 7.5 | **Pilot projects** | If at all possible, run a pilot project covering a few programs or a small system, and staff the project with the best people, not just those that are available. Make any financial commitment to licence the reverse engineering method dependent on a successful pilot. |

# 8 Future trends in reverse engineering

The predecessor to reverse engineering and re-engineering, albeit with more limited objectives, can be considered to be the conversion industry. Several software houses specialised in porting software systems from one environment to another, often as a consequence of an organisation's purchase of a new mainframe.

There is currently a considerable amount of technology in which conversion is achieved by the hardware, or by emulation software which requires little or no change to the application software.

In the longer term, compatibility between application systems may be provided by distributed computing and use of the open systems 'Posix' family of international standards (ISO 9945 and related standards). But there are few other formal international standards in this area: most are de facto standards based on achieving compatibility with a commonly used proprietary specification.

Some proprietary systems software is widely available, for example, IBM's proprietary CICS software is available on a range of compatible mainframes, medium sized systems and OS/2 personal machines. The X windows system is supported by a number of medium to small and personal machines.

Additionally, it is becoming possible to emulate operating environments within other environments. The result is that it is not always necessary to port an application that was formally supported by the emulated environment.

However, the need for reverse engineering remains to support application software development. Those services are still relatively immature and unstructured, and few are widely available yet. This situation can be expected to change as a result of:

- improving technology for reverse engineering

- better understood management techniques which will identify reverse engineering as low risk

- the fact that, increasingly, business process analysis is undertaken after application software has been installed to support the business function, and may have been running for many years. The analysis must, therefore, take account of the existing computer system rather than be simply the analysis of non-automated manual procedures

- reducing cost of hardware, but not software.

In the short term, the most promising technological developments are the emergence of repositories. These may store not just code, but also all software associated items, for example, documentation and test suites. This development should also lead on to the establishment of integrated methods and tool sets operating on the repository. Such toolsets will address the full complement of the code and the associated elements. It is expected that tools will emerge to help load repositories initially, and then transfer information between repositories.

# Chapter 8
## Future trends in reverse engineering

Looking at the longer term, research in reverse engineering has until recently received less priority than other areas of software engineering. There are recent indications that much more work is now being undertaken to find solutions to reverse engineering problems. To a certain extent, research is fragmented and lacks an overall coherent approach – the problem is now seen as a serious one – but there are a diversity of solutions proposed. As these solutions are explored, it is expected that some will be rejected and there will be more general agreement in the field on worthwhile solutions.

## 8.1 Technical developments

### 8.1.1 Formal methods

In forward engineering, there is the start of some limited use of certain mathematical techniques for specifying a software system and then deriving an executable code version from the specification. These efforts are taking place outside academic research laboratories and involve languages such as Z or VDM.

The use of such methods as formal transformations in reverse engineering is starting to move out of the ivory towers of academia into commercially available methods and tools. Formal methods have three advantages:

1. they are exact

2. they have the ability to represent and transform between software at different levels of abstraction; for example, code, design, specification

3. they can be used in subsequent forward re-engineering.

### 8.1.2 Cognitive methods

A substantial part of reverse engineering (and maintenance) is concerned with system understanding. Multi-disciplinary research into software engineering, with both psychology and artificial intelligence, may lead to better tools and methods based on the requirements of this research. In a closely related field, continually improving human-computer interfaces are becoming available. These are likely to be of benefit for addressing large scale software systems.

### 8.1.3 Tools and methods

Currently, many reverse and re-engineering tools are not offered as an integrated toolset, and have not been developed with any particular method in mind. This is an indication of an immature discipline. No de facto reverse engineering standards exist at present, although some de jure standards are beginning to emerge for some underlying components and data interchange formats, for example IRDS, PCTE and CDIF.

However, we can anticipate integrated methods evolving for reverse engineering, using repositories integrated with wider maintenance and forward engineering methods. These methods will result in much improved tools, including some that will work at higher levels than code.

### 8.1.4 Reuse

Software reuse and reverse and re-engineering are closely linked. Better understanding of reuse (described in section 3.8) and of the software processes in which it can take place, are likely to be beneficial to reverse engineering.

# Chapter 8
## Future trends in reverse engineering

## 8.2 Management issues

### 8.2.1 Quality

Quality management and process maturity are applicable to reverse and re-engineering, and there is wide opportunity to adopt these techniques. As reverse engineering processes reach greater maturity, metrics will be developed to assess the effectiveness of their application.

### 8.2.2 Case studies

Currently, there are relatively few reverse engineering case studies written up in the public domain in which the authors have abstracted, from a particular set of experiences, wider lessons that apply beyond their own situation. It is hoped and expected that in time this will be remedied.

### 8.2.3 Risk reduction and costing

As further experience with reverse engineering is gained, it should increasingly be seen that, applied in appropriate circumstances, the present low risks of the technique will reduce to an even lower level. Chapter 4 gives details of the benefits to be expected. Reverse engineering is becoming easier to plan and monitor: the costs and benefits are becoming predictable; for example, the level of maintainability is becoming more easily measured. Of course, if reverse engineering is applied in inappropriate circumstances the risk of not achieving the benefits is increased.

### 8.2.4 Personnel

There is some indication that maintenance in general and reverse engineering in particular, are no longer being seen as 'dead-end' unmotivating jobs. As software engineering matures, software maintenance will be seen as a product support/development activity much more in line with the continuous product development seen in other engineering disciplines.

# Annex A: Examples of reverse engineering projects

A.1 **Using reverse engineering to redocument a system.**

An application used in the insurance industry was key to the process of calculating the premiums due. Because of the nature of the application it had undergone many changes to reflect revisions in the way that cover was offered and premiums calculated. An additional factor was the changes in the way data was input and extracted since the original design.

The application was required to run for a further number of years, but because of the difficulty in training maintenance staff, it was decided to redocument the system.

Although it may have been possible to complete the task without it, the use of reverse engineering tools gave the following benefits:

- accuracy of the analysis

- completeness of the references to data items and files

- speed - the time to complete the task was significantly reduced

- unnecessary user involvement was kept at a minimum

- the documentation was captured on magnetic media for future ease of use.

A.2 **Downsizing**

Because of changes in ownership, an organisation was running its applications on a mainframe which was owned by another company not now part of the same group. The costs of the service were likely to increase significantly in the future.

The functionality of the applications met the organisation's requirements and so, rather than redevelop or buy a new package for a mid-range machine, the possibility of moving the mainframe applications to a mid-range machine was considered.

Reverse engineering was used extensively both to analyse the code and to judge its suitability for transfer and, after the decision had been taken to proceed, in the transfer itself.

A.3 **Understanding the code**

The maintenance of a suite of applications had been contracted out and was being run to a service level agreement based on response times for both corrective and perfective maintenance. A change to the VAT rate had to be completed in a short time. Reverse engineering tools used by staff who had no detailed knowledge of the code or its history revealed:

- some programs only catered for a two digit code (it was now NN.N)

- despite reading a variable VAT parameter, the calculation was hard coded to use a fixed rate in some areas

- there were other dubious constructions which may have caused problems.

Once identified, the problems were corrected and the changes implemented within the contracted service level and quality standards.

A.4 **Standardising existing code**

The benefits of using data dictionaries and repositories were being realised by an organisation using the latest standards in the development of application code. This only served to highlight that, although the older code was reliable and would be required for a number of years, perfective maintenance was becoming more difficult.

It was decided to adopt the standard data names throughout the older systems. Reverse engineering was used to identify the nature and use of the original (programmer chosen) names, and also the use and construction of common modules. These were then changed to the new standards and the references built into the repositories.

Annex A
Examples of reverse engineering projects

**A.5  Quality of new code**  Reverse engineering can be used to check the quality of new software code. It is doubtful if such an activity would justify acquiring a tool specifically for this purpose. However, one company that already has a reverse engineering tool uses it to check the quality of software developed both by outside suppliers and in-house development teams.

# Annex B: Examples of cost improvement through reverse engineering

Most cost/benefit improvement through the use of reverse engineering techniques arises from two sources:

- increasing the productivity of the support and maintenance staff

- extending the life of a system and delaying the costs associated with any future redevelopment.

**B.1  Productivity increases**

Section 4.1.2 outlines two factors which have a major influence on productivity:

- the time taken for maintenance support staff to become proficient is on average 6 months and can be over a year

- up to 50% of a maintenance programmer's time can be spent in 'figuring out' the code (understanding the code in detail and the implications of the way in which it is structured) before making any changes.

Reverse engineering can provide additional information about the build up and detail of a system. This may reduce the time required for staff to become proficient in navigating around the system, or the time taken to understand the code associated with a particular modification.

For teams of over 5 staff, the direct saving can be easily identified as the reduction in staff required to support the system. Each organisation will show different savings but these can be in the order of 25%.

Significant savings may be made when small teams (3-4 staff) dedicated to supporting a single system can be merged and, by means of the information provided from reverse engineering, support several systems. The original team size is often determined by the need to retain resilience for sickness and holidays, rather than from the resources needed to complete the current workload. In these cases savings of up to 50% can be achieved.

Including apportioned overhead and management costs, the saving is likely to be in the area of 2.5 to 3 times salary per person per annum.

### B.2 Extending the life of a system

The following examples make many assumptions about costs of development, training, maintenance and support. Whilst they are not specific to a particular development, they are broadly in line with computer industry averages, serving to illustrate the mechanisms whereby savings can be made by using reverse engineering to extend the life of a system.

Consider the investment in a system (exclusive of hardware) to run for 5 years, of the order of:

| | |
|---|---:|
| Cost of the initial development | £1,000,000 |
| User design/development input +15% | £150,000 |
| Training/installation +25% | £250,000 |
| Modifications/support at 12% pa | £600,000 |
| | |
| Total cost over 5 years | £2,000,000 |
| Cost per year | £400,000 |

If this system is then redeveloped to include the modifications made to the original system, and is then run and supported for a further 5 years, then:

| | |
|---|---:|
| Development cost | £1,400,000 |
| User design/development | £210,000 |
| Training/installation | £367,000 |
| Modifications and support | £840,000 |
| | |
| Total cost over the next 5 years | £2,817,000 |
| Cost per year | £563,400 |

# Annex B
## Examples of cost improvement through reverse engineering

If the original system's life can be extended by 2 years as a result of a £100,000 reverse engineering exercise, then:

| | |
|---|---:|
| Cost of reverse engineering | £100,000 |
| Cost of support for years 6 & 7 | £240,000 |
| Total cost for years 6 & 7 | £340,000 |
| Savings against cost for redeveloped system in years 6 & 7 | £786,000 |

If the original system can be reverse engineered and then re-engineered to extend its life for a further 5 years, then:

| | |
|---|---:|
| Cost of the engineering work | £400,000 |
| Cost of support | £600,000 |
| Total cost for years 6 to 10 | £1,000,000 |
| Savings over the 5 years | £1,817,000 |

# Annex C: Reasons for reverse engineering

There are a number of reasons why an organisation may undertake the reverse engineering of a software system.

The following tables show the issues in checklist form and have been categorised into quality, management and technical lists.

These checklists are an extract from the REDO Compendium (see the Bibliography for details).

C.1  **Quality issues**

The likely benefits of reverse engineering in improving the quality of application software are:

1   simplification of complex software

2   identification and removal of errors

3   removal of side-effects of the implemented system that come from unplanned state changes

4   improved coding quality

5   the opportunity to undertake a major design repair activity, because the original design (and, therefore, the implementation) may have be erroneous

6   the production of up-to-date, consistent and helpful documentation that is relevant to maintainers

7   the establishment of a proper strategy and test suite for regression testing and verification, validation and system testing

8   meeting the need for improved performance in the software system, such as reducing the storage space needed, or increasing the speed of execution

9   bringing existing application software into a modern software engineering development environment, to be used in a manner consistent with other practices within the organisation

10  allowing the design to be validated as part of a financial auditing process

11  allowing quality and maintenance inspection to be introduced into the quality auditing of the software

12  incorporating the reverse engineering project within the quality management system (QMS).

## C.2 Management issues

Chief management issues for reverse engineering are:

1   the enforcement of organisational, national or international program and programming standards

2   enabling the adoption of better techniques for the management of software maintenance; in particular, the planning, monitoring and control of development using appropriate metrics

3   addressing legal aspects. It may be necessary in the future to undertake reverse engineering because of legal arguments about the copyright of software

4   enabling an organisation to produce information such as 'impact of change' reports and audit trails, and to cost them more effectively.

## C.3 Technical issues

The technical implications of reverse engineering are:

1. it allows major changes to be implemented. The structure, documentation and quality of the existing software may be so poor that it is not feasible to implement a major change without a reverse engineering exercise. This would express the application function at a higher level of abstraction that could, using modern techniques and tools, be implemented and maintained

2. it rediscovers the underlying business model implicit in the software

3. it rediscovers and records the design of the system

4. it rediscovers and records the requirement specification of the system; in both this and the previous case, it may be appropriate and useful to provide more than one view

5. it can port the application to a new environment or re-express it in a new programming language

6. it helps to establish and support a reuse policy for code, designs, specifications and processes

7. it enables the redesign of exception handling in the system, including the possible introduction of fault tolerance

8. it recognises that the software maintenance practice is itself evolving, therefore there is a need to review the software

9. it discovers, as far as possible, the original requirements of the software, because they were never stated or even known

10 it recovers and records high-level information about the system including:

- the system structure, in terms of its components and their interrelationships expressed by interfaces

- the system's functionality in terms of what operations are performed on what components

- the dynamic behaviour of the system, in understanding how input is transformed to output

- the system rationale. Design involves deciding between a number of alternatives at each design step. It may be necessary to record which alternatives were chosen and why the decision was made the way it was

- the system's construction. It may be necessary to determine which modules, documentation and test suites are stored in which files in the filestore or database.

11 it rediscovers which tools or hardware platforms were used to generate the system

12 it updates the source code so that it can be compiled and linked by the tools that are currently supported, rather than those which were used when the software was built

13 it rediscovers the contents of the software release (in terms of configuration management) possibly required due to a disaster

14 it replaces old design concepts (which at the time were well implemented) by modern techniques

15 it facilitates the interfacing of the system to a new or modified kernel in a layered architecture.

# Annex D: Reverse engineering methods, tool types and tool selection

Reverse engineering is a broad term, and the need to use reverse engineering can originate from several sources. As a consequence the functionality of the software products described as reverse engineering tools can vary. The classifications listed below are based on the output requirement from the tool. No attempt is made to distinguish between those tools which provide full reverse engineering functionality and those which provide only partial functionality.

**D.1  Code analysis tools**

There are many tools available for code analysis. Most are static tools and only analyse the code as presented. They do not provide facilities for the user to update the code. Any changes to the code have to be made using the normal edit and update procedures, after which the code is re-analysed by the tool.

Typical types of tool cover:

- quality control checking
- code and data validation
- program understanding
- impact analysis
- code and design metrics.

The majority of the tools analyse on a program by program basis. However, there are some which will also analyse the complete system, including the job control language (JCL) statements and the files accessed.

**D.2 Documentation tools** — The purpose of these tools is to produce (or assist in producing) documentation from the existing program code and job control statements. For this reason documentation tools are often similar to code analysis tools and have, in addition, enhanced printing and graphics output. Some will provide the output in a standard word processing format for future amendment.

**D.3 Conversion tools** — These have many of the attributes of the code analysis and documentation tools. They enable the user to identify incompatibilities between the current and proposed environments. However, they do not usually identify inefficient code, or code which may be difficult to understand and maintain. Some tools will automatically substitute new code for such functions as file access.

Many of these tools produce converted code only, and do not provide the intermediate levels of information such as structures and data flows, which are needed for future maintenance and support.

These tools are offered by:

- software vendors, as stand alone software products

- hardware suppliers, to encourage users to convert the existing systems to use their hardware

- software houses offering an overall conversion service, but not selling the tool as a separate product.

At present most of these tools operate only on a program by program basis.

# Annex D
## Reverse engineering methods, tool types and tool selection

**D.4  Development tools**  Many of the latest development tools offer comprehensive code extraction, analysis and search facilities. Two factors limit the use of these functions for reverse engineering type activities:

- they are primarily development tools and their functionality is always directed towards development support

- they are only available for the later releases of the standard languages such as COBOL, and considerable effort may be required to upgrade the older systems to be compatible with these tools.

**D.5  Restructuring tools**  These tools analyse the program code and enable the programmer to manipulate and restructure the code, usually one program at a time. Restructuring tools are more sophisticated than code analysis tools, and have many of the functions contained within conversion tools, such as code substitution.

# Annex E: Contracting out of reverse engineering services

Reverse engineering may impact contracted out services in two areas:

- a project to reverse engineer a system or particular piece of code

- processing and support services which may need to use reverse engineering for the benefit of the service provider or the customer.

The Information Systems Engineering Library volume *Management of Software Maintenance* discusses the contracting out of software maintenance services in detail.

There are a number of points to be considered which apply specifically to contracted out reverse engineering, as indicated below.

## E.1 Projects

For a reverse engineering project:

- has the scope been defined in a way that can cater for the exploratory nature of the task?

- have the technical risks been evaluated and can the task be completed as one phase?

- are the tools used generally available or are they specific to the particular supplier; that is, can the task be carried out elsewhere or in-house?

- how will the output be presented and in what form?

Reverse Engineering - An Overview

- how many of the supplier's staff are capable of completing the task?

- who will own the reverse engineered system and the documentation?

- has a pilot project been considered?

E.2 **As part of processing or support services**

The need to use reverse engineering as part of processing or support services arises from the requirement to understand the application system and its code, prior to making improvements to benefit the service.

These improvements are likely to be:

- an increase in system reliability, by removal of 'hot spots'

- a reduction in processing time or IT capacity

- functional modifications

- modifications to enable the code to run on a new version of the operating system, database or new hardware.

The key points to consider are:

- whose responsibility is it to identify the need to reverse engineer?

- will any changes made as part of the project result in 'lock-in' to that supplier?

- will the organisation's normal current use of standards, methods and procedures be followed?

- how will the success criteria be set and measured?

# Annex E
## Contracting out of reverse engineering services

- who pays for the reverse engineering project?

- who benefits from the results?

- who takes the risk?

- can the supplier reverse engineer without informing the client?

- who owns the information produced from the reverse engineering?

Further information and advice on availability and capacity planning issues may be found in the IT Infrastructure Library volumes: *Availability Management* and *Capacity Management*.

Reverse Engineering - An Overview

# Bibliography

**Appraisal and Evaluation Library**

The Appraisal and Evaluation Library volumes are available from HMSO Books (Dept A), Freepost, Norwich, NR3 1BR, or telephone 071 873 9090, fax 071 873 8200.

The following volumes are referenced in this publication:

- Overview and Procedures
  ISBN 0 11 330534 6

- CASE Tools
  ISBN: 0 11 330609 1

- IT Infrastructure Support Tools
  ISBN: 0 11 330586 9

**Information Systems Guides**

The Information Systems Guides are available from John Wiley and Sons Ltd, Baffins Lane, Chichester, PO19 1UD.

The following volumes are referenced in this publication:

- CCTA IS Guide volume A5: A Project Manager's Guide
  ISBN: 0 471 92525 X

**Information Systems Engineering Library**

The Information Systems Engineering Library volumes are available from HMSO Books (Dept A), Freepost, Norwich, NR3 1BR, or telephone 071 873 9090, fax 071 873 8200.

The following volumes are referenced in this publication:

- CASE and the Issues for IS Management
  ISBN: 0 11 330594 X

- Estimating with Mark II Function Point Analysis
  ISBN: 0 11 330578 8

- Improving the Maintainability of Software
  ISBN: 0 11 330585 0

- Management of Software Maintenance
  ISBN: 0 11 330584 2

**IT Infrastructure Library**

The IT Infrastructure Library volumes are available from HMSO Books (Dept A), Freepost, Norwich, NR3 1BR, or telephone 071 873 9090, fax 071 873 8200.

The following volumes are either referenced in this publication or are likely to be of interest to readers:

- Availability Management
  ISBN: 0 11 330551 6

- Capacity Management
  ISBN: 0 11 330538 7

- Change Management
  ISBN: 0 11 330525 7

- Configuration Management
  ISBN: 0 11 330530 3

- Help Desk
  ISBN: 0 11 330522 2

- Problem Management
  ISBN: 0 11 330527 3

- Software Control and Distribution
  ISBN: 0 11 330537 0

- Software Lifecycle Support
  ISBN: 0 11 330559 1

- Testing an IT Service for Operational Use
  ISBN: 0 11 330560 5

# Bibliography

**Programme and Project Management Library**

The Programme and Project Management Library volumes are available from HMSO Books (Dept A), Freepost, Norwich, NR3 1BR, or telephone 071 873 9090, fax 071 873 8200.

The following volumes are referenced in this publication:

- Guide to Programme Management
  ISBN: 0 11 330600 8

- PRINCE - A Management Outline
  ISBN: 0 11 330599 0

**User Interface booklets**

The User Interface booklets are available from the CCTA Library, CCTA, Riverwalk House, 157-161 Millbank, London, SW1P 4RT, or telephone 071 217 3331.

- User Interface: The Issues
  1990
  ISBN: 0 946683 36 0

- User Interface: Style Guide Issues
  1991
  ISBN: 0 946683 54 9

- User Interface: Style Migration Issues
  1991
  ISBN: 0 946683 57 3

Other publications

Reverse Engineering and Design Recovery: a Taxonomy
Chikofsky and Cross
IEEE Journal of Software, Volume 7 Number 1, 1990

Reverse Engineering Markets, Methods and Tools
R Rock-Evans and Hales
Ovum 1990
ISBN 0 903969 64 5

An Overview of Maintenance and Reverse Engineering
The REDO Compendium: K H Bennett
ISBN 0 471936 07 3

Economics of Software Re-engineering
Software Maintenance and Research Practice, Volume 3 (1991): H Sneed
John Wiley

Methods and Tools: Spikes
Spikes Cavell & Co

IEEE Standard Glossary of Software Engineering Terminology
Institute of Electrical and Electronic Engineers
Standards Board
ISBN 1 55937 067 X

# Glossary

| | |
|---|---|
| adaptive maintenance | A change made to application software to adapt it for a change of the supporting environment, or network or hardware platform. |
| applications | Information systems which support a specific business area. |
| backup (noun) | A system, component, file or procedure used to replace, or help restore, a primary item in the event of failure or externally caused disaster. |
| backup (verb) | The action of creating a system, component, file or procedure that will be the backup for a failed primary item. |
| business case | The financial and management justification for IT proposals. |
| business process | A course of action which progresses all, or a particular element, of the business to its conclusion. |
| call out | The request that programmers and support staff should return to the computer centre, or office, outside normal working hours to investigate and correct processing irregularities. |
| CASE | Computer Aided Software Engineering. The use of computerised aids in the design and development of software systems to increase the productivity of developers, and the quality of their products. |
| change control | The control of changes to established systems. This process includes: identification of a need for change, evaluation of the cost of change and its impact, obtaining authority (or not) to make the change, and co-ordination of the change with other developments during implementation. |
| checkpoint | A point in a computer program at which the program state, status or results are checked and recorded. |

| | |
|---|---|
| COBOL | A programming language widely used for business applications since the mid 1960s. |
| compiler | A computer program that translates programs expressed in a high order language into their machine language equivalents. |
| corrective maintenance | The correction of processing, performance or implementation problems in application software. |
| database | A collection of interrelated data stored together in one or more computerised files. |
| debug | To detect, locate and correct faults in a computer program. |
| design recovery | Recreating the architecture, components, interfaces and other characteristics of the system from the operational code and instructions. |
| document image processing | The use of a combination of hardware and software tools to scan documents and represent their images in digitised form which can then be stored and manipulated in computers and transmitted across networks. |
| electronic data interchange (EDI) | A set of standards and interfaces used for the exchange of commercial trading information across computer networks such as the placing of orders and deliveries for goods. |
| emulator | A device, computer program or system that accepts input in the format of another type of system and mimics the behaviour of that other system. |
| escrow | Copies of information (such as the source code for a package) that are not normally released by the prime vendor, but are held by a third party. Such copies may be released to the client of the third party under specific conditions; particularly if the prime vendor ceases to trade and can no longer support the package. |

# Glossary

| | |
|---|---|
| facilities management (FM) | The provision of the management, operation and support of an organisation's IT systems and/or computers and/or networks by an external source at agreed service levels. The service is generally provided for a set time at an agreed cost. |
| forward engineering | See software engineering. |
| Help Desk | A centralised point for the initial handling of queries and requests for assistance from users. |
| 'hot-spot' | A particular program or procedure that regularly fails, or does not meet performance and quality criteria. |
| IEEE | Institution of Electrical and Electronic Engineers. |
| impact analysis | An examination of the effect of an error, or proposed system change, on the system's operation and on the businesses dependent on it. |
| installation (verb) | The loading and integration of software into a computer to form an operational system. |
| installation (noun) | One or several computers and supporting services operating singly or together, usually on a single site. The term is more commonly used to describe mainframe and mid-range machines. |
| IS | Information Systems. |
| IT | Information Technology. |
| job control language (JCL) | A language used to identify a sequence of jobs, describe their requirements to an operating system and control their execution. |
| kernel | A software module that encapsulates an elementary function, or functions, of a software system. |
| mainframe | The term often used to describe large, powerful multiprocessing computers. They usually require siting in air conditioned and air filtered rooms with a separate, smoothed power supply. |

| | |
|---|---|
| maintainability | The ease with which a software system can be corrected when errors or deficiencies occur. Also, the ease with which the system can be expanded, modified or contracted to satisfy new requirements. |
| memory | Part of a computer's internal storage used for holding active programs during execution and their associated active data. |
| metric | A quantitative measure of the degree to which a system, component or process possesses a given attribute. |
| mid-range | A computer usually having less power than a mainframe, but with a significant multiprocessing capability. It can often be sited in a normal office environment using office power supplies. |
| migration | Modifying a set of computer programs to run on different hardware and/or operating software. |
| non-functional requirements | Features of a unit of application software that do not directly serve that software's business function, but are required to allow it to run in a particular environment, and to interact with other software. |
| operating environment | The hardware configuration and the associated systems software and utilities, which control and are used by the application software. |
| operating software | See operating environment. |
| package (application) | A set of programs and associated documentation written and distributed to provide a standard solution for a particular business need across a range of organisations or industries. |
| parameter | A variable, external to the program code, whose value can be changed on entry to, or exit from, the code without the need to make an internal software change. |
| perfective maintenance | Any modification or enhancement of the existing functionality, or performance of, application software. |
| platform | See operating environment. |

| | |
|---|---|
| portability | The ease with which a software system can be transferred from one hardware/software configuration to another. |
| preventive maintenance | Action taken to make subsequent maintenance of application software more efficient and reliable. |
| process maturity | The condition reached by a process whereby its defined purpose, function and application are generally understood and agreed. |
| recovery | The restoration of a system, program, database or other system resource to a state in which it can perform required functions. |
| redocumentation | recreating documentation from examination of the software code and/or other operational procedures. |
| re-engineering | The examination and alteration of software to reconstitute it in a new form and the subsequent implementation of the new form. |
| regression testing | The testing of an application software system, or subsystem, following a change to any part of that system, or a system with which it interacts. The tests are those devised to validate the functionality of the pre-existing system. |
| reliability | The ability of a system or component to perform its required functions under stated conditions for a specified period of time. |
| report writer | A computer program that can extract data from various files, process that data and present the results in a user-chosen format. The choice of data, processing and presentation format are decided, at invocation, by user-given parameters, without any need to change internal code. |
| repository | A library providing permanent archival storage for software and related documentation. |
| restart | To cause a computer program to resume execution after a failure, using status and results recorded at a checkpoint or backup. |

| | |
|---|---|
| restructuring | The reshaping of software for improved maintainability by making it:<br><br>• easier to understand and to change<br><br>• less susceptible to errors when later changes are made. |
| reverse engineering | The process of analysing a software subject to identify the system's components and their inter-relationships, and create representations of the system in another form, or at higher levels of abstraction.<br><br>These representations make the software subject more amenable to enquiry, analysis, reuse and documentation. Reverse engineering may require the use of a repository, or the generation of information in an appropriate form and notation for re-engineering into a new system using CASE tools. |
| reuse | The incorporation of a software module or other work product in more than one computer program or software system. |
| SDM | Structured Design Method |
| software engineering | The science and art of specifying, designing, implementing and evolving - with economy, timeliness and elegance - programs, documentation and operating procedures that make computers useful to people. |
| software library | A controlled collection of software and related documentation designed to aid in software development, use or maintenance. |
| software maintenance | Any modification of a software product after delivery to correct faults, to improve performance or other attributes, or to adapt the product to a changed environment. (IEEE Standard P1219.) |

# Glossary

| | |
|---|---|
| Structured Systems Analysis and Design Method (SSADM) | SSADM is a non-proprietary and publicly available method which provides a structured set of procedural, technical and documentation standards designed specifically for analysing business needs and undertaking software development. |
| tools | Software packages which help the developers or maintainers of software systems to improve the productivity or quality of their work. |
| unattended operations | A computer installation which does not require continuous on-site supervision and monitoring. Sometimes referred to as a dark centre, because the lights are switched off in the computer room. |
| Z | A language for the mathematical representation of program function specifications. |

# Reverse Engineering - An Overview